LITTLE LIBRARY

Space

Christopher Maynard

Kingfisher Books

NEW YORK

Contents

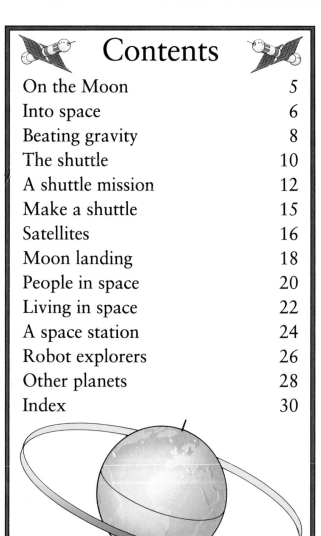

On the Moon

P eople have dreamed about going to the Moon for hundreds of years. It was a dream that finally came true about 25 years ago when two American astronauts set foot on the Moon for the first time.

Since then, more astronauts have visited the Moon. Robot spacecraft have also traveled to the other planets to find out more about them.

Into Space

T he only way a spacecraft can get into space is on top of a rocket. Rockets are used to travel into space because no plane has yet been built which has the power, or the speed, to climb that far above the Earth. Also, rocket engines can be made to work in space where there is no air.

THE FRONTIER WITH SPACE

All around the Earth is a thick layer of air called the atmosphere. The higher up you go, the thinner it gets. About 60 miles (100 km) above the Earth is the frontier with space. Here, there is no air or wind. It is very hot in the sun and very cold in the shade. There is no blue sky, only black space.

Beating gravity

The Earth pulls everything to it with a force called gravity. The only way to escape the force of gravity is to use a huge rocket that can push through it at a speed of about 25,000 mph (40,000 km/h).

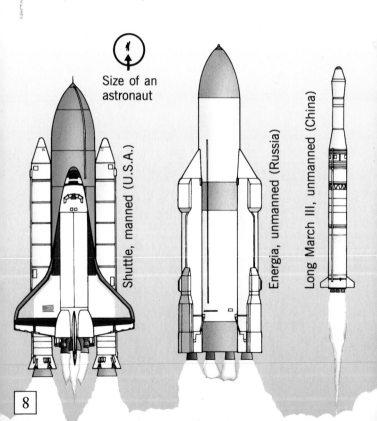

Size of an astronaut

Shuttle, manned (U.S.A.)

Energia, unmanned (Russia)

Long March III, unmanned (China)

▷ A rocket is made up of two or three parts. As each part uses up its fuel, it falls away. Only the last part goes into space where it may be used to launch a satellite. Some rockets carry spacecraft that take people into space. They are called manned spacecraft. Others are unmanned – they don't carry any people.

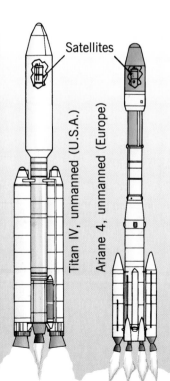

Satellites

Titan IV, unmanned (U.S.A.)

Ariane 4, unmanned (Europe)

◁ A number of countries have powerful rockets. But only the U.S. shuttle can be used again. This is because it takes off like a rocket but after it returns to Earth's atmosphere, it glides down to land on a runway like a plane.

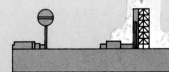

The shuttle

Ordinary rockets are expensive to build and can't be used again. The shuttle solves this problem. It's a spaceplane that can fly into space and back many times. After each flight the orbiter and the rocket boosters are re-used. Only the fuel tank is lost. It falls away when all the fuel has been used.

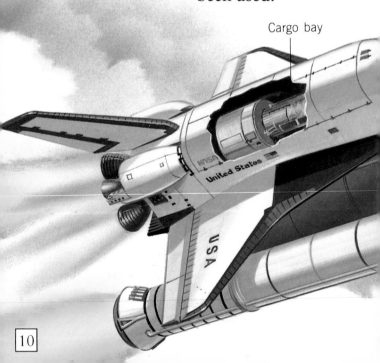

Cargo bay

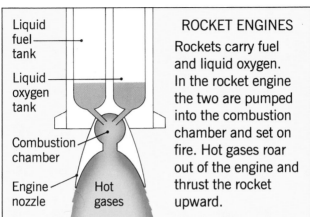

ROCKET ENGINES

Liquid fuel tank

Liquid oxygen tank

Combustion chamber

Engine nozzle

Hot gases

Rockets carry fuel and liquid oxygen. In the rocket engine the two are pumped into the combustion chamber and set on fire. Hot gases roar out of the engine and thrust the rocket upward.

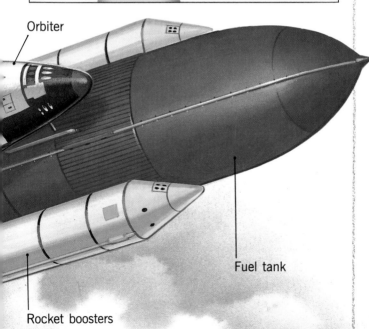

Orbiter

Fuel tank

Rocket boosters

11

A shuttle mission

The crew's most important job during a shuttle mission is to launch satellites. The cargo bay, which is as big as a school bus, can fit several satellites inside it at a time.

On other missions, the shuttle is used as a laboratory where the scientists do tests and experiments.

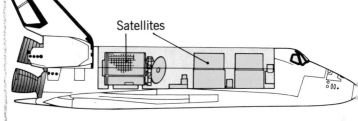

Satellites

On some missions, the cargo bay of the shuttle is loaded with 3 or 4 satellites. On others, all kinds of scientific equipment is carried in a space laboratory called Spacelab.

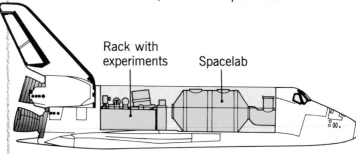

Rack with experiments

Spacelab

A SHUTTLE FLIGHT

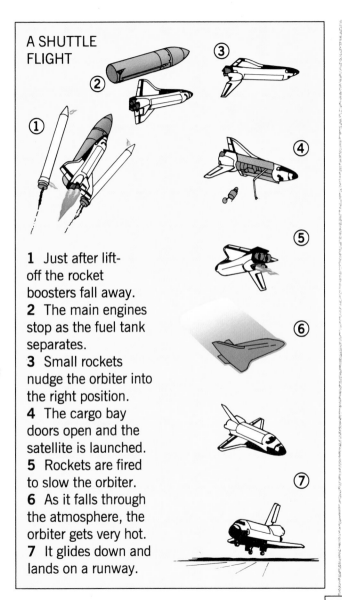

1 Just after lift-off the rocket boosters fall away.
2 The main engines stop as the fuel tank separates.
3 Small rockets nudge the orbiter into the right position.
4 The cargo bay doors open and the satellite is launched.
5 Rockets are fired to slow the orbiter.
6 As it falls through the atmosphere, the orbiter gets very hot.
7 It glides down and lands on a runway.

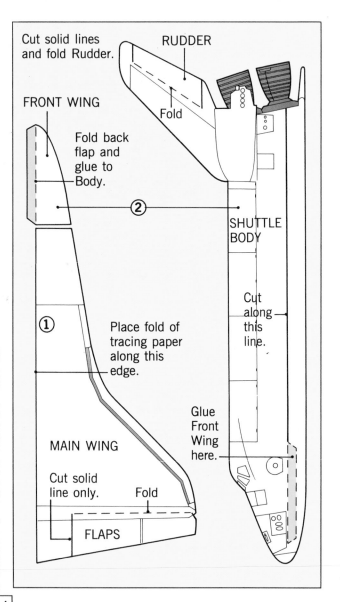

Cut solid lines and fold Rudder.

RUDDER

Fold

FRONT WING

Fold back flap and glue to Body.

② SHUTTLE BODY

Cut along this line.

① Place fold of tracing paper along this edge.

MAIN WING

Glue Front Wing here.

Cut solid line only.

Fold

FLAPS

14

Make a shuttle

The orbiter is the main part of the shuttle. It carries the crew into space and returns them safely to Earth.

Try making your own shuttle. You will need some tracing paper, cardboard, glue, tape, scissors, and paper clips.

1 Fold the tracing paper in half. Place the fold on the Main Wing. Trace around it. Open up the paper and trace again to get both wings.
2 Trace the Front Wing twice and Shuttle Body once. Glue them onto cardboard and cut out.
3 Cut the line on the Shuttle Body. Cut and fold the Rudder.
4 Cut and fold the Main Wing Flaps. Slot the Wing into the Body and tape it in place.
5 Fold and glue the Front Wing Flaps to the Body. Put paper clips on the nose for balance.

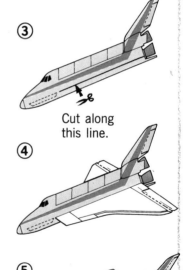

③

Cut along this line.

④

⑤

Satellites

W hen a satellite is launched into space, it circles the Earth again and again along the same path. This path is called an orbit.

Satellites are used in many ways. Some take pictures of cloud movements which help the weather forecasters. Others pick up television programs or telephone conversations from one continent and send them on to another. Satellites can also be used by ships to help them find a clear way through ice.

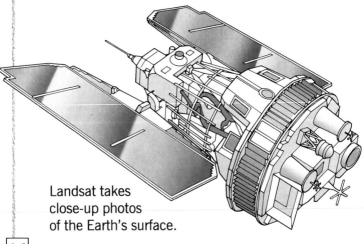

Landsat takes close-up photos of the Earth's surface.

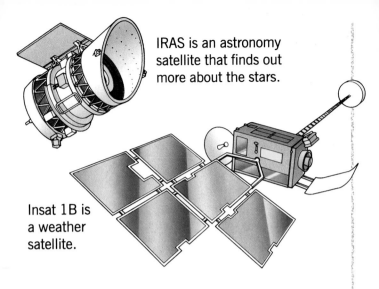

IRAS is an astronomy satellite that finds out more about the stars.

Insat 1B is a weather satellite.

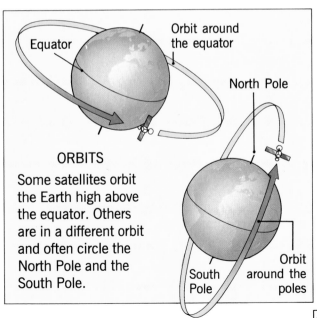

Equator

Orbit around the equator

North Pole

ORBITS

Some satellites orbit the Earth high above the equator. Others are in a different orbit and often circle the North Pole and the South Pole.

South Pole

Orbit around the poles

Moon landing

The first astronauts to reach the Moon landed in July 1969. Two years later another team brought a moon buggy, the most expensive "car" ever built. It cost 60 million dollars.

Lunar module

Moon buggy

TRIP TO THE MOON

The astronauts were launched in a rocket the size of a skyscraper and came back in a capsule as small as a small car.

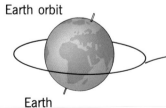

Earth orbit

Earth

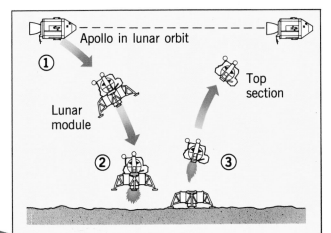

Apollo in lunar orbit

① Lunar module

② ③ Top section

1 U.S. astronauts went to the Moon in an Apollo spacecraft.
2 They climbed into the lunar module to land on the surface.
3 Only the top half of the lunar module took off to rejoin the Apollo spacecraft.

It took nearly 3 days to travel to the Moon.

The journey was about 240,000 miles (385,000 km), which is almost 10 times around the world.

Lunar orbit

Moon

People in space

The men and women who go into space are called "astronauts" by Americans and "cosmonauts" by the Russians. But there is no difference. They do the same job.

To prepare themselves, they train in dummy spacecraft and in special tanks so they can learn what it feels like to move around and work in space.

△ Astronauts train in water tanks to get the feel of what it will be like working in space.

▷ This is what the astronauts had to wear during their walks on the Moon.

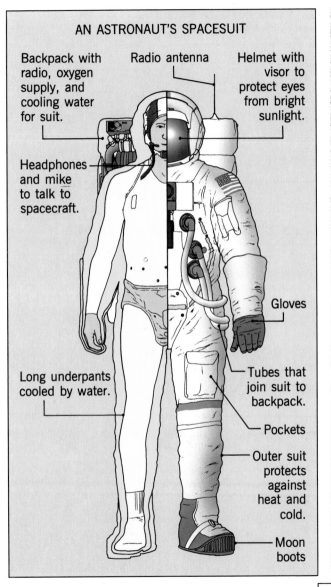

AN ASTRONAUT'S SPACESUIT

Backpack with radio, oxygen supply, and cooling water for suit.

Radio antenna

Helmet with visor to protect eyes from bright sunlight.

Headphones and mike to talk to spacecraft.

Long underpants cooled by water.

Gloves

Tubes that join suit to backpack.

Pockets

Outer suit protects against heat and cold.

Moon boots

Living in space

The first thing astronauts notice in space is that they have no weight. Once they unbuckle their seat belts they just float around in the air.

Astronauts discover they can stand on the walls of the spacecraft just as well as on the floor. They can even sleep on the ceiling if they want to.

In the Russian space station Mir, shown below, crews often stay for months. From time to time, spacecraft from Earth fly up with fresh supplies.

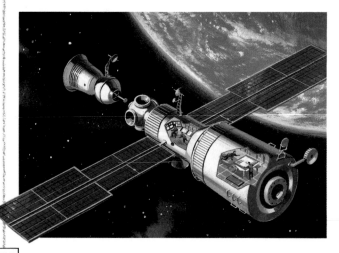

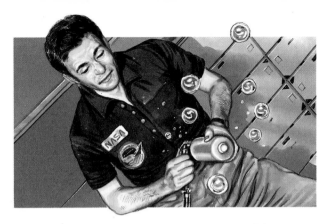

△ Drinking and washing is difficult in space because any liquid forms little balls that just float around. When astronauts want to have a drink, they have to suck through a straw or squirt the liquid into their mouth.

▽ It's impossible to walk around in space because there is nothing to keep your feet down. With every step, you just float upward. The astronauts use special suction shoes when they want to stand in one place.

23

A space station

T he Russian space station Mir has a lot more room to work in than a crowded spacecraft. This means that the cosmonauts can stay in space for longer and they can carry out many more experiments.

BUILD A SPACE STATION

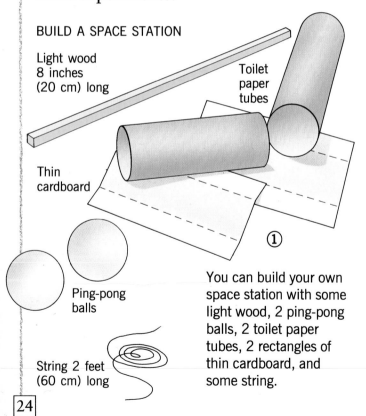

Light wood
8 inches
(20 cm) long

Toilet
paper
tubes

Thin
cardboard

①

Ping-pong
balls

String 2 feet
(60 cm) long

You can build your own space station with some light wood, 2 ping-pong balls, 2 toilet paper tubes, 2 rectangles of thin cardboard, and some string.

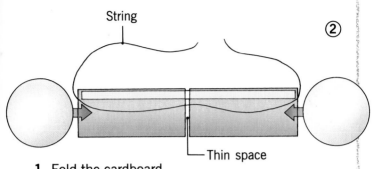

String

②

Thin space

1 Fold the cardboard along the dotted lines to make solar panels.
2 Thread the string through the two tubes. Glue the wood inside the tubes leaving a thin space in between the tubes. Glue the balls onto the ends.
3 Slot the solar panels into the gap between the two tubes and glue them in place.

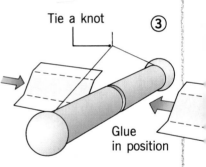

Tie a knot

③

Glue in position

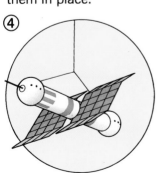

④

4 Paint your space station in bright colors to make it look like a real one. Then tie a knot in the string and find somewhere to hang your space station.

Robot explorers

S ome spacecraft are sent out to explore the planets. They are called probes. Sometimes they may travel for years before they get close enough to a distant planet to send back information about it.

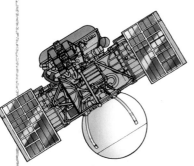

Venera probe

The Russian probe, Venera, found that Venus had a temperature of over 850°F (450°C), which is hot enough to melt lead. The weight of the air, or air pressure, was 100 times more than Earth's.

AIR PRESSURE POWER

You can't feel the weight of air pressure because it presses equally around you all the time. But on the planet Venus, an Earth person would crumple just like the bottle in this experiment.

1 This test shows just how powerful air pressure is. Take a soft plastic bottle with a screw top and ask an adult to fill it with hot water. Then ask them to pour out the water. Screw the top on quickly.

2 Run cold water over the bottle. The bottle collapses because the air inside it cools and shrinks as the pressure becomes less than the air pressure outside.

Other planets

In 1976, two Viking probes landed on Mars to find out more about its surface. They sent back photos of a bare, rocky planet with no signs of life on it at all. Other probes have flown past Venus, Mercury, Jupiter, Saturn, Uranus, and Neptune.

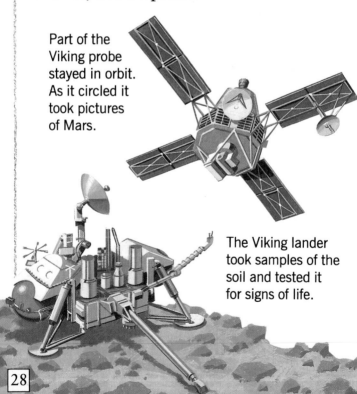

Part of the Viking probe stayed in orbit. As it circled it took pictures of Mars.

The Viking lander took samples of the soil and tested it for signs of life.

MESSAGES IN SPACE

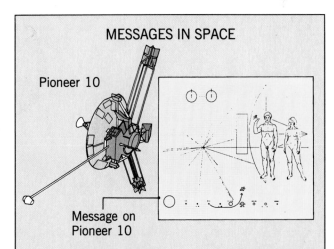

Pioneer 10

Message on
Pioneer 10

Two American space
probes, Pioneer 10 and
11, are now heading
for the stars. On board,
they carry a picture to
show where the probes
came from and who
sent them. If another
form of life ever found
the picture it would
learn that Earth exists.

Make a list of the
things you would put
in a message to outer
space. What kind of
objects would tell
someone most about
life here?

Fill a tin
box with your
space message.

TIMES

Menu

Index

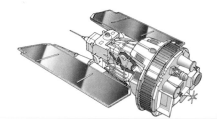